Rafael Novaes
Antonio Vitor Machado
Patricio B Maracaja

Pombal Slaughterhouse Effluent Diagnosis, Treatment and Reuse

Rafael Novaes
Antonio Vitor Machado
Patricio B Maracaja

Pombal Slaughterhouse Effluent Diagnosis, Treatment and Reuse

Diagnosis of effluents generated at the public slaughterhouse in the municipality of Pombal-PB, Sustainable Treatment and Reuse project

Imprint
Any brand names and product names mentioned in this book are subject to trademark, brand or patent protection and are trademarks or registered trademarks of their respective holders. The use of brand names, product names, common names, trade names, product descriptions etc. even without a particular marking in this work is in no way to be construed to mean that such names may be regarded as unrestricted in respect of trademark and brand protection legislation and could thus be used by anyone.

Cover image: www.ingimage.com

This book is a translation from the original published under ISBN 978-3-330-73996-3.

Publisher:
Sciencia Scripts
is a trademark of
Dodo Books Indian Ocean Ltd. and OmniScriptum S.R.L publishing group

120 High Road, East Finchley, London, N2 9ED, United Kingdom
Str. Armeneasca 28/1, office 1, Chisinau MD-2012, Republic of Moldova, Europe
Printed at: see last page
ISBN: 978-620-7-88547-3

Table of contents:

RAFAEL DA SILVA NOVAES

DIAGNOSIS OF THE EFFLUENTS GENERATED AT THE PUBLIC

SLAUGHTERHOUSE IN THE

MUNICIPALITY OF POMBAL-PB, PROJECT FOR TREATMENT AND

SUSTAINABLE

REUSE

SUMMARY

Among the agro-industrial activities carried out in Brazil, the slaughtering of cattle, pigs and poultry stands out as one of the most important national economic activities on the market. Slaughterhouses are an economic activity for the industrialization or processing of agricultural products of animal origin. Although at different technological levels, and in general, the main impacts generated by these activities are related to the large volume of drinking water consumption, the generation of large flows of industrial effluents, the treatment of effluents and the environmental impacts caused by the disposal of this waste. Reducing the volume of agro-industrial waste, cleaner production, rational use and reuse of water are topics of great interest today given the scarcity of water resources in the country. With this in mind, the main objective of this research was to diagnose the effluent from the treatment system of the municipal slaughterhouse in Pombal - PB, characterizing it physically and chemically and assessing its efficiency in terms of the parameters required by current legislation. As a field of study, the effluent treatment system of the municipal slaughterhouse in Pombal - PB was evaluated by defining sample collection points for determining efficiency parameters, and the following analyses were carried out: Hydrogen Potential (pH), Total Phosphorus (FT), Total Nitrogen (NT), Chemical Oxygen Demand (COD), Biochemical Oxygen Demand (BOD5), Sedimentable Solids (SSed), Turbidity, Oils and Grease in accordance with the methodologies described by the competent bodies regulating this activity. The results of the parameters analyzed were compared with the legislation based on resolutions CONSEMA 128/06 and CONAMA 357/05. The physiochemical *characterization* revealed high indices for some variables, demonstrating that it is in disagreement with the parameters stipulated by current Brazilian legislation, thus violating the environmental laws of this activity. In view of this situation, it is recommended that changes be made to the layout of the treatment system in order to meet the requirements of current legislation, with the adoption of techniques for its reuse once it has been treated in the slaughterhouse itself.

Keywords: Slaughterhouse, effluent treatment, reuse.

Chapter 1

1 INTRODUCTION

Among the agro-industrial activities carried out in Brazil, the slaughter of cattle, pigs and poultry stands out. It is one of the national economic activities that can be considered one of the most important in the market, and Brazil is one of the largest producers and exporters of meat in the world. According to the Brazilian Association of Meat Exporting Industries - ABIEC (2014), the estimate for 2015 regarding Brazilian beef production is approximately 14,000 tons, which is surpassed only by the United States, in terms of beef exports, Brazil ranks first in the world, exporting more than 2,650 tons/year.

Slaughterhouses are one of the economic activities involved in industrializing or processing agricultural products of animal origin. Although at different technological levels, and in general, the main impacts generated by these activities are related to the large volume of drinking water consumption and, consequently, the generation of large flows of industrial effluents (ABIPECS, 2015).

The operational process in slaughterhouses is concentrated on the production of fresh meat and its cuts, while the other by-products (blood, skin, hair and guts) require processing and cleaning. This wastewater contains blood, lard, organic or inorganic solids, as well as other salts added during processing operations, making up a large amount of dissolved and suspended material, which is classified as having great polluting potential (MATOS, 2015).

In Brazil, industries and agriculture stand out in terms of drinking water consumption. It is known that for a long time water was considered to be an infinite resource, but this theory has been falling apart as a result of changes in its availability and especially its quality due to the inappropriate use of this resource, which makes drinking water an increasingly scarce resource today.

According to the National Water Resources Policy, any industrial activity must treat its effluents, but the lack of public policies aimed at monitoring water resources often makes it impossible to comply with current legislation. What's more, of the few industries that do carry out treatment, not all reuse this precious resource.

Water reuse is a significant area of potential application in Brazil. The use of treated effluents in urban areas and agriculture, for non-potable purposes, to meet artificial aquifer recharge and industrial demands, is a powerful instrument for restoring the balance between water supply and demand in various Brazilian regions. However, water reuse needs to be institutionalized and regulated in order to promote it, so that in practice appropriate technological principles are developed that are economically viable, socially acceptable and, above all, environmentally correct (HESPANHOL, 2013). As a result, there is a need not only to treat water, but also to reuse it in some way in industrial processes, thus providing better management of water resources in the country.

Due to the need to conserve water resources, it is extremely important to develop studies aimed at the planned reuse of effluents resulting from the treatment of industrial wastewater, as a way of reducing the consumption of drinking water in activities that do not require this type of water, thus generating a reduction in the use of good quality water sources, promoting a rational environmental use of this precious natural resource.

In this way, this study was carried out by analyzing the effluent from the public slaughterhouse in the municipality of Pombal-PB, comparing the results obtained with the standards required and established by current Brazilian legislation, suggesting a sustainable and viable treatment project for the reuse of this effluent in the municipal slaughterhouse itself; Alternatives will be proposed for the treatment of the existing effluent, in order to optimize the process and meet the environmental standards required by current legislation, also proposing ways of applying this effluent for environmentally correct reuse.

Chapter 2

2 - Objectives

2.1 - General objective

To diagnose the effluent from the treatment system of the municipal slaughterhouse of Pombal - PB, characterizing it physically and chemically in order to propose alternatives in the treatment process, with a view to adapting it to the parameters of current legislation and indicating alternatives for local reuse.

2.2 - Specific objectives

a) Physico-chemical characterization of the effluent from the municipal slaughterhouse in Pombal-PB, with a view to its reuse.

b) Compare the results of the effluent analysis with the standards required by current national legislation for this industrial activity.

c) Designing a treatment plant, suggesting modifications to the existing local plant in order to optimize the process and meet legislative requirements.

d) To suggest ways of reusing the effluent within the slaughterhouse itself, with a view to preserving water resources, using the treated water for better purposes and reducing the environmental impact of this activity.

Chapter 3

3 BIOGRAPHICAL REVIEW

Among the natural resources available to human beings, water is one of the most important and indispensable for maintaining life on earth. However, the increasing use of this resource has resulted in problems not only in its scarcity, but also in changes to its quality.

Planet Earth is particularly covered in water, more precisely 2/3 of the earth's surface is made up of this renewable natural resource that is indispensable for the life of any living being. However, not all water is suitable for more noble uses, such as drinking, given that approximately 97.5% of water is salty and is contained in the seas and oceans,5% of water is salty and is contained in the seas and oceans, approximately 2.493% of water is fresh or drinkable, and as if that weren't enough, access to this small portion is still restricted, since only 0.007% of it is available to humans in rivers, lakes and the atmosphere, which can be consumed (NOVAES, 2014).

For a long time, water was considered to be an infinite resource, but this theory has become outdated due to changes in its availability and, above all, its quality due to the inadequate use of this resource, which means that every day drinking water is reduced in volume, thus causing numerous problems ranging from political order, for example, conflicts between nations and misery in the regions most in need of water, to the problem of environmental health (NOVAES, 2014).

The inadequate distribution of water has a negative impact on any nation. In Brazil, for example, which is the country with the highest percentage of drinking water sources in the world, the country still has major problems of scarcity due to the inadequate national distribution of water, as well as containing considerable quantities of polluted rivers, affecting the health of the population and thus hindering the use of these waters for more noble purposes (SALES, 2013).

When it comes to the inadequate internal distribution of water here in Brazil, it can be seen that the Northeast is the region with the biggest problems on a national scale, with only 3.30% of the country's water resources (Table 1).

Table 1. Distribution of water, area and population by region in Brazil

Region	Resource	Surface	Populated
North	68,50%	45,30%	6,98%
Midwest	15,70%	18,80%	6,41%
South	6,50%	6,80%	15,05%
South East	6,00%	10,80%	42,65%
North East	3,30%	18,30%	28,91%

Ponte: Secretaria de Recurso" Hídrico" do Meio Ambiente a|n<d Jornal do Medio Vale, Nov. 1, 2003, p.ll

Source: Paludo et al (2013)

In addition to having only 3.30% resources, its population is the second largest in the country at 28.91%. Therefore, with few water resources, a considerable number of inhabitants, a crystalline geology and irregular rainfall patterns and times, understanding the problem of water in this region is crucial for its sustainable development (FAPPI, 2015).

Faced with this scenario, characterized by a region of water scarcity, a strategy was needed, at first, to "combat drought" and, later, to adapt, to "live with the semi-arid region".

Some of the strategies used to "combat drought" include the construction of large, medium and small dams in the region. One of the main water storage strategies in the Brazilian Northeast is agudagem, which consists of building a dam in order to reserve water for a longer period of time, making it available in times of drought. This is one of the oldest practices in the semi-arid region. The first dams were built in the 19th century and expanded in the 1960s due to the experience of previous droughts (NOVAES, 2014).

The Northeast's dams are considered preponderant not only for human supply, but also for energy generation, which is the case, for example, with the Sobradinho dam, located in Bahia, which is responsible for a large part of the electricity consumed in the region.

When it comes to dams in the Northeast, the main disadvantage is the high evapotranspiration, which is the main limiting factor for building new dams in the semi-arid region. On the other hand, the quality of water from dams is highly seasonal and subject to negative changes in its composition, which can make it unsuitable for more noble uses, such as drinking and animal watering.

Groundwater would be an excellent alternative for human supply in the semi-arid region, because it is protected from pollutants and evaporation. However, their potential is quite limited due to the predominance of crystalline bedrock in much of the northeastern subsoil, making them mostly unsuitable for consumption, and they are not always protected from contamination, due to the difficulties in implementing environmental management in this region, mainly due to the lack of public policies.

The use of desalination plants in the northeastern semi-arid region is a practice that has been widely advocated by state governments, but the high cost of installing these processes has been a limitation to acquiring them, and it must also be taken into account that the waste from this process must be sent to appropriate locations (SALES, 2013).

Among the main water storage strategies in the semi-arid region, rural cisterns are perhaps the most important. This is because they provide a considerable volume of quality water for human consumption when operated properly, offering good quality water throughout the year. In addition, cistern construction ends up generating sources of income for countless parents in the Northeast (BRANCO, 2006).

In the context of water scarcity, when it comes to water availability in the present and future, it has become increasingly necessary to manage water, which, if not implemented as soon as possible, will only increase the environmental impacts. It is therefore up to public authorities to provide better management of this finite resource, which is essential for life on earth.

3.1.1 Reuse of water resources

For a long time, mankind considered water to be an inexhaustible resource. There is no shortage of examples of fresh water scarcity, seen in the lowering of groundwater levels, the "shrinking of lakes" and the drying up of marshes. On the other hand, worldwide concern is growing about rational use, the need to control losses and waste, and the reuse of water. This includes the use of sanitary sewage for various purposes: water reuse, providing relief in demand and preserving the water supply for multiple uses, nutrient recycling, meaning savings in the production of fertilizers and animal feed and, above all, a reduction in the discharge of sewage into receiving bodies (VON SPERLING, 2005; BLUM, 2003).

Water reuse is not a new concept and has been practiced all over the world for many years. There are reports of its practice in Ancient Greece, with the disposal of sewage and its use in irrigation. However, the growing demand for water has made the planned reuse of water a current issue of great importance, and water reuse should be considered as part of a broader activity that is the rational or efficient use of water, which also includes the control of losses and waste, and the minimization of effluent production and water consumption. From this point of view, treated sewage plays a fundamental role in the planning and sustainable management of water resources as a substitute for the use of water for agricultural purposes, irrigation and other purposes, because by freeing up good quality water sources for public supply and other priority uses, the reuse of sewage contributes to the conservation of resources and adds an economic dimension to the planning of water resources (BLUM, 2003).

According to Filho and Mancuso (2003), part of this context is the need to implement a water reuse system, which can occur spontaneously in nature, in the hydrological cycle, or through human actions, whether planned or not. The technique of planned reuse consists of using water more than once, reusing it for the same or a different purpose after it has undergone treatment. According to the same author, reuse can be:

> Unplanned indirect: occurs when water, used in some human activity, is discharged into the

environment and used again downstream, in its diluted form, in an unintentional and unplanned way, subject to the natural stresses of the hydrological cycle;

> Planned indirect: occurs when the effluents, after being treated, are discharged in a planned manner into surface or groundwater bodies, to be used downstream in a controlled manner to meet some beneficial use;

> Planned direct: this occurs when the effluent, once treated, is taken directly from its point of discharge to the place of reuse and is not discharged into the environment;

> Water recycling: the internal reuse of water in a given process, before it is discharged into a general treatment system or disposal site.

According to Asano (1998), reuse water from treatment plants, where domestic and industrial sewage can undergo numerous treatment processes, can be used for a wide variety of purposes:

> Landscape irrigation: parks, cemeteries, highway medians, university campuses, greenbelts, residential lawns, monument cleaning;

> Irrigation in agriculture: planting fodder crops, fibrous and grain plants, food plants, ornamental plant nurseries, frost protection;

> Industrial uses: cooling, boiler feed, process water, washing handles and tanks, power generation;

> Aquifer recharge: for marine intrusion control, subsoil settlement control;

> Non-potable urban uses: landscape irrigation, firefighting, toilet flushing, sewage system unblocking, air conditioning systems, vehicle washing, street and bus stop washing, etc;

> Environmental purposes: increasing flow in watercourses, application in marshes, wetlands, fishing industries;

> Various uses: aquaculture, civil construction, marine intrusion control, water control for animal use and dust control.

Brackish water, which is second quality and not as salty as sea water, and agricultural drainage water can also be used as reuse water. It is very important to emphasize that effective reuse requires measures such as: evaluation of treatment systems, definition of use criteria, adequate planning and monitoring, resulting water quality and control of the environmental impacts and benefits resulting from the practice. Agriculture is a sector where reuse needs to be applied urgently, as 80% of the water consumed in the world is used in this sector and in Brazil this percentage is 70% for irrigation. Treated effluent can be used on certain crops and the adoption of methods such as the furrow process also favors the conservation of drinking water (ASANO, 1998).

3.1.2 Water Resources Legislation

Since the beginning of the last century, Brazil has been producing legislation and policies that gradually seek to consolidate a way of valuing its water resources.

It wasn't until the 1980s, with the creation of Federal Law No. 6.938/1981, which sets out the National Environmental Policy, and Article 225 of the 1988 Federal Constitution, which defines the environment as a common good that must be preserved for future generations, that water came to be understood as a finite good that is indispensable to the quality of life (OLIVEIRA, 2004; BRASIL, 1988).

Water management in Brazil has undergone a period of great progress since the late 1980s. In 1997, with Law 9.433, which instituted the National Water Resources Policy and the creation of the National Water Resources Management System - SINGREH, emphasis was placed on the sustainable use of water.

With the creation of the National Water Agency - ANA, Law 9.984/2000, directly linked to the Ministry of the Environment, with administrative and financial autonomy, responsible for implementing action instruments to control and regulate the use of resources and the release of pollutants that affect the environment. This law is based on some basic principles, such as: adoption of the river basin as the planning unit; guarantee of the multiple use of water resources; recognition of water as a finite, vulnerable resource and an asset of economic value, thus instituting charging for its use and provision for decentralized and participatory management, with the shift of decision-making power to the local and regional hierarchical levels of government, and the participation of users, organized civil society, NGOs and other agents through basin committees (BRASIL, 2000).

3.1.2.1 CONAMA No. 357 of March 17, 2005

Resolution No. 357 of March 17, 2005 of the National Environment Council (CONAMA) provides for the classification of water bodies and gives environmental guidelines for their classification, as well as establishing the conditions and standards both for these water bodies and for the discharge of effluents (CONAMA, 2005).

These guidelines are based on water quality standards related to the predominant uses of the water body, which establish individual limits for each substance in each class. In this way, the classification of watercourses aims to adapt the current and intended restrictive uses to a desired level of quality, in such a way as to make them compatible with anthropogenic activities, while maintaining the aquatic ecological balance (SANTOS, 2009).

In this sense, according to Articles 4, 5 and 6 of this resolution, fresh, brackish and saline waters are classified as: special class, class 1, class 2 or class 3, according to the quality required for their predominant uses (CONAMA, 2005).

Table 2: Parameters to be monitored for surface water

Parameters	Standard	CONAMA Resolution 357 (2005)*
Ph		6-9
Turbidity (NTU)		100
Nitrites (mg/L ')		1
Nitrate (mg.L)[4]		10

Parameter	Method	Value
Color (mg.L[-1] Pt)		75
Odor	APHA(1998)	N.R
COD (O_2 mg.L)[1]		N.R.
BOD5 (mg.L)[1]		5
Dissolved Oxygen (mg.L)[1]		NO less than 5
Thermotolerant coliforms (NMP.lOOml)[4]		1000 NMP.lOOml [1]

Source: Adapted from CONAMA (2005).

3.1.2.2 CONSEMA No. 128 of November 24, 2006

Resolution No. 128 of November 24, 2006 of the State Environment Council (CONSEMA) provides for the setting of Effluent Emission Standards for emission sources that discharge their effluents into surface waters in the state of Rio Grande do Sul. For liquid effluents from polluting sources, except domestic liquid effluents, it is established that the emission standards will vary according to the flow ranges of the polluting source.

Table 3 shows the parameters to be monitored in surface water

Table 3: Parameters to be monitored in surface water

Parameters	Resolution a" 128 CONSEMA (2006)
Ph	6-9
Temperature (°C)	<40

Total Nitrogen (mg.L^{-1})[1]	<20
Phosphorus (mg.L^{-1})[1]	<3
Color (mg.L^{-1} Pt)	It must not change the color of the receiving water body.
Odor	Free of unpleasant odors
COD (O$_2$ mg.L^{-1})[1]	<300
BOD$_5$ (mg.L^{-1})[1]	<80
Suspended Solids (mg.L^{-1})[1]	<100
Thermotolerant coliforms or *Escherichia coli* (MPN 100ml^{-1})[1]	10000 MPN 100ml^{-1}
Oils and fats: vegetable and animal	<30
(mg.L^{-1})[1]	
Sedimentable solids (mg.L^{-1})	<1

Source: Adapted from CONSEMA (2006).

3.1.2.3 CNRH Resolution No. 54/2005

Resolution 54 of the National Council of Water Resources (CNRH), of November 28, 2005, establishes modalities, guidelines and general criteria that regulate and encourage the practice of direct non-potable water reuse (Art. 1). Among its criteria are the considerations that no water of good quality should be used in activities that tolerate water of inferior quality, since water resources must be conserved for public supply or other more demanding uses; that water reuse is a practice for rationalizing and conserving water resources, in accordance with the principles set out in Agenda 21; that the rising costs of water treatment are due to the degradation of water sources and, for this reason, the practice of reuse is a factor in reducing the discharge of pollutants into receiving bodies; and that reuse contributes to protecting the environment and public health.

Among its main proposals, the Resolution included:

a) The bodies that make up the National Water Resources Management System should establish regulatory and incentive instruments for the various types of reuse according to their effects on water bodies (Art. 4).

b) The producer, distributor or user of reuse water who, as a result of the reuse activity, changes the quantity or quality of the interventions in the water body contained in the terms of the current grant, must request the competent authority to rectify the grant of the right to use water resources in order to adjust it to the changes (Art. 5).

c) Any activity involving water reuse must be reported to the water resources management body, the environmental body and the public health body for the purposes of registering the grantee (producer, distributor or user), including the place of origin and destination of the reused water, as well as its purpose and the daily volume involved in the activity (Art. 9 and its sections).

Although this Resolution makes some very important demands regarding the granting of the right to use reuse water, and regarding information for registration, it has delegated the encouragement of reuse to other SINGERH bodies when it itself could have decided on more incentive or cogent criteria. Despite the fact that the requirement to grant the right to use water for reuse has a certain character of administrative police power, since it is conditional on the classification of the water according to CONAMA 357/05, (DANTAS, *et al* 2009).

3.1.3 Water quality indices

Water quality indices reflect the level of healthiness, the behavior of the ecosystem, as well as indicating the condition of the aquatic environment and act as complementary tools for analyzing the water quality of a river. In addition, they can give an idea of the trend in water quality over time, allowing comparisons to be made between different watercourses (LEMOS, 2003). The main objective of these indices is to determine the potential for ecosystem dysfunction and to enable a better understanding of the sources of contamination and management decisions (ONGLEY, 2000 *apud* NUNES, 2008).

According to Ott (1978) *apud* Nunes (2008), there are three basic types of water quality indices: (1) indices based on expert opinion; (2) indices based on statistical methods; and (3) biological indices.

As such, these indices can have various applications, such as: distributing resources and determining priorities; comparing environmental conditions in different geographical areas; determining compliance with environmental legislation; analyzing trends; evaluating changes in environmental quality; providing information to the public; scientific research; identifying water quality problems that require special studies in stretches of river; among others (NUNES, 2008).

The main advantages of water quality indices are that they are easy to communicate to the non-technical public, that they have a higher status than individual parameters and that they represent an average of several variables in a single number, combining different units of measurement in a single unit. However, its main disadvantage is the loss of information on the individual variables and the interaction between them. Although the index provides an integrated assessment, it will never replace a detailed assessment of water quality in a given river basin (CETESB, 2010).

In the 1970s, the *Water Quality Index* (WQI) was developed by the *National Sanitation Foundation* (NSF) in the United States. It was based on the Rand Corporation's Delphi technique, through surveys of various environmental experts (PINTO, 2007; NUNES, 2008).

Based on this study, the Minas Gerais Institute of Water Management (IGAM) developed the IQA. To this end, nine relevant parameters were considered for assessing the quality of Brazilian waters: dissolved oxygen (DO), fecal coliforms, pH, biochemical oxygen demand (BOD), nitrate, total phosphate, temperature, turbidity and total solids (PINTO, 2007).

The nine parameters of the WQI better identify the water quality of streams contaminated by domestic effluents, with a view to using them to characterize water intended for general public supply (COELHO, 2008 *apud* SANTOS, 2009).

The selection of these parameters was based on the opinions of water quality experts. Initially, 35 parameters were proposed for evaluation, with their respective weights and condition, according to a *rating* scale. For the nine parameters selected, water quality variation curves were established according to their status, as well as their corresponding relative weight (VON SPERLING, 2005).

3.1.4 Slaughterhouse activity

Many companies use more water than necessary, usually due to a lack of control over the volumes being used. Cleaning operations are mainly responsible for the high water consumption in slaughterhouses, as the floors in the process areas must be washed and sanitized at least once a day. The water consumed in the cleaning and washing of carcasses represents more than 80% of the water used in the effluent generated (ENVIROWISE, 2000 *apud* KRIEGER 2007).

The total amount of water used per animal varies between slaughterhouses and depends on the *layout, the* type of animal slaughtered, the slaughter techniques and the degree of automation. Different units are used to express water consumption, which makes it difficult to compare consumption between companies. In slaughterhouses with meat and fat processing, consumption is between 3 and 6 m3/t of animal slaughtered, after the implementation of clean technologies, and in Brazil it is between 0.4 and 3 m3/pig (SENAI, 2003).

Approximately half of the water used in slaughterhouses, Figure 1, is heated from 40 °C to 60 °C. In pig slaughterhouses, hot water from shaving operations contains a large amount of hair (ENVIROWISE, 2000 *apud* KRIEGER, 2007).

FIGURE 1: Distribution of water consumption in different process areas in a large pig slaughterhouse.

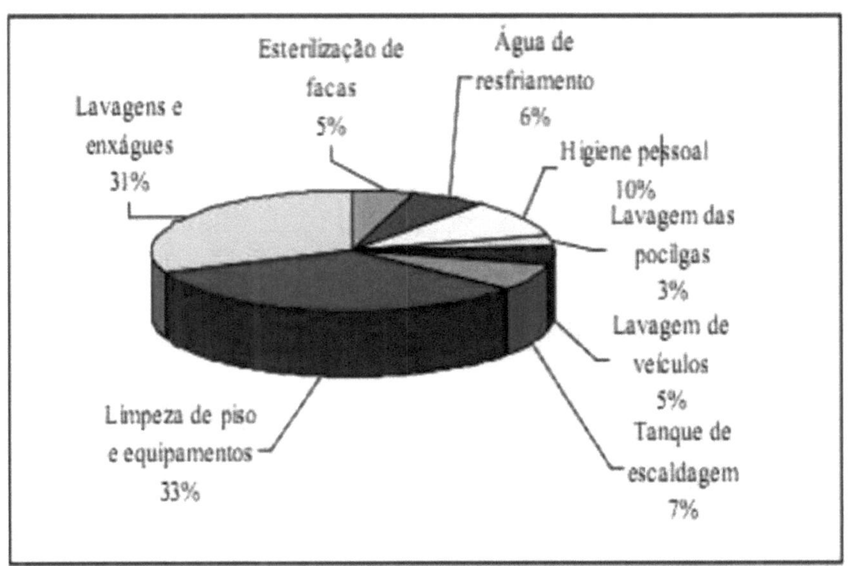

3.1.4.1 Treatment for slaughterhouse effluents

Water consumption in industries is influenced by various factors such as: production capacity, climatic conditions in the region (which will determine the quantities of water consumed in heat exchange processes), water availability, production method, age of the facility (newer industries use more modern technologies, with equipment that is less susceptible to downtime and maintenance), operating practices and the culture of the company and the local community (MIERZWA, 2002).

Slaughterhouses use large quantities of water due to strict hygiene standards. Water is used for watering animals and washing pens, washing trucks, scalding, washing viscera and carcasses, transporting products and waste, cleaning and sterilizing knives, equipment and floors, feeding boilers and cooling compressors and condensers (KRIEGER, 2007).

According to UNEP (2000) *apud* KRIEGER (2007), 80 to 95% of the water consumed in slaughterhouses becomes effluent, which contains high levels of organic matter due to the presence of manure, fat and blood. The effluent can also contain concentrations of salts (sodium), phosphates and nitrates from manure and stomach contents.

Blood is the main contributor to the effluent's organic load, with a total COD of approximately 375,000 mg/L. It is also the largest contributor of nitrogen, and it is estimated that between 15% and 20% of blood is lost as effluent (CHILE, 1998).

Nitrogen occurs mainly in the form of ammonia, due to the breakdown of protein material into amino acids. However, as the nature of the ammonia species present depends on pH, nitrogen concentrations in slaughterhouses are commonly expressed as total nitrogen. Proteins and grease are important components of the organic load present in washing water, along with other substances such as heparin, bile salts, carbohydrates, detergents and disinfectants. Noteworthy is the high content of pathogenic microorganisms, such as *Salmonella* and *Shiguella* bacteria, parasite eggs and amoeba cysts, and pesticide residues from the treatment and feeding of animals (KRIEGER, 2015).

3.1.4.2 Characteristics of slaughterhouse treatment systems

Most of the technologies used to treat effluents for reuse are the same as those used in water and effluent treatment systems, but in certain cases additional treatment processes are required to remove specific contaminants and to inactivate and remove pathogenic microorganisms (METCALF and EDDY, 2003).

The main technologies used in wastewater treatment systems for reuse and their purposes are described below (LEVINE *et al*, 2002):

a) For solid/liquid separation

- sedimentation: for the removal of particles larger than 30 pm. It is also used as primary treatment after secondary treatment;

- filtration: for removing particles larger than 3 pm, used after sedimentation (conventional treatment) or after coagulation/flocculation.

b) Biological treatment (secondary)

- aerobic biological treatment: for the removal of dissolved or suspended organic matter;

- Oxidation ponds: to reduce suspended solids, BOD, pathogenic bacteria and ammonia;

- biological nutrient removal: for nitrogen and phosphorus reduction;

- disinfection: to protect public health by removing pathogenic organisms;

c) tertiary or advanced treatment

- activated carbon: for the removal of hydrophobic organic compounds;

- *air strpping*: for the removal of ammoniacal nitrogen and some organic volatiles;

- ion exchange: for removing cations such as calcium, magnesium, iron, ammonia and anions such as nitrate;

- chemical coagulation and precipitation: for the formation of phosphorus precipitates and flocculation of particles to be removed by sedimentation and filtration;

- lime treatment: used to reduce the water's fouling potential, precipitate phosphorus and change the pH;

- membrane filtration: for removing particles and microorganisms from water;

- reverse osmosis: for removing dissolved salts and minerals from the solution, and is also effective for removing pathogens.

Primary treatment is the initial stage of an effluent treatment process, with an expected efficiency of 50% for removing suspended solids, 25 to 50% BOD, 10 to
20% organic nitrogen and approximately 10% phosphorus. Efficiency can be increased with the addition of coagulants and flocculants. For most industrial reuse applications, primary treatment is insufficient to meet the necessary quality standards (LEVINE *et al,* 2002).

The secondary treatment system provides adequate removal of biodegradable organic matter and is often supplemented by filtration for additional particle removal and disinfection, making it suitable for reuse in many industrial processes. Tertiary treatment is applied after biological treatment (LEVINE *et al,* 2002).

According to CETESB's Environmental Technical Guide for Meatpacking Plants (2008), in order to minimize the environmental impacts of their industrial liquid effluents and comply with local environmental legislation, meatpacking plants must treat these effluents. This treatment may vary from company to company, but a typical treatment system in the sector has the following stages:

- Initial separation or segregation of liquid effluents into two main lines: "green" line, which mainly receives the effluents generated in the reception of the animals, in the pens, in the conduction to the slaughterhouse/"syringe", in the truck washing areas, in the bucharia and in the triparia; and "red" line, whose main contributors are the effluents generated in the slaughterhouse, in the processing of meat and offal, including the deboning/cutting and greasing operations, if they take place in the industrial unit;

- Primary treatment: for the removal of coarse, suspended, settleable and floatable solids, mainly by physical-mechanical action. The following equipment is generally used grids, sieves and manure/dunghouses (the latter, on the "green" line, in units with slaughtering), to remove coarse solids; then grease traps (with or without aeration) and/or flotators, to remove grease and other flotable solids; then sedimenters, sieves (static, rotary or vibrating) and flotators (dissolved air or electroflotation), to remove settleable, suspended and emulsified solids - finer or smaller solids. Primary treatment is carried out for the "green" line and the "red" line separately;

- Equalization: carried out in a tank of suitably defined volume and configuration, with a constant outflow and precautions to minimize the sedimentation of any suspended solids, by means of mixing devices. It can absorb significant variations in the flow rates and pollutant loads of the liquid effluents to be treated, attenuating peak loads for the treatment plant. This facilitates and optimizes the operation of the plant as a whole, helping to achieve the desired final parameters in the treated liquid effluent. In slaughterhouses, equalization is carried out by bringing together the effluents from the "green" and "red" lines, which, after their primary treatment and equalization, go on to continue treatment;

- Secondary treatment: for the removal of colloidal, dissolved and emulsified solids, mainly by biological action, due to the biodegradable nature of the remaining content of primary treatment effluents. At this stage, there is an emphasis on stabilization ponds, especially anaerobic ones. The following are possibilities for anaerobic biological processes: anaerobic lagoons (widely used), contact anaerobic processes, anaerobic filters and upflow anaerobic digesters. With regard to aerobic biological processes, we can have film aerobic processes (biological filters and biodiscs) and dispersed biomass aerobic processes (activated sludge - conventional and prolonged aeration, which includes oxidation ditches).

It is also quite common to see the use of photosynthetic lagoons in the sequence of treatment with anaerobic lagoons. There can also be anaerobic treatment followed by aerobic treatment;

- Tertiary treatment (if necessary, depending on local environmental requirements): carried out as a final "polishing" of liquid effluents from secondary treatment, promoting additional removal of solids, nutrients (nitrogen, phosphorus) and pathogenic organisms. Associated nitrification-denitrification systems, filters and biological or physical-chemical systems (e.g. use of coagulants for phosphorus removal) can be used. When there is a grease factory attached to the slaughterhouse, there can be variations, such as individualized primary treatment and subsequent mixing of its primary effluents in the unit's general equalization tank; mixing the raw effluent from the grease factory with the effluents from the "red" line, at the entrance to its primary treatment, among others (SCARASSATI, 2013).

3.1.5 Qualitative effluent parameters

The quality of water and effluents can be represented by various parameters, which reflect the main physical, chemical and biological characteristics. The physical characteristics are mostly associated with the solids present in the water. The chemical characteristics can be interpreted through two classifications: organic or inorganic matter. The biological characteristics are linked to the microorganisms present in the water (VON SPERLING, 2005).

Color is a characteristic of the substances dissolved in water (BRAGA, 2002). According to Libánio (2005), it results from the reflection of light off particles with a diameter of less than 1.0 pm, as well as the presence of metallic compounds or the leaching of effluents into the receiving water body.

Natural waters have a color that varies between zero and 200 UTNs (Turbidity Units), as above this would be swamp or marsh water, with high levels of dissolved organic matter. Coloration below 10 UTNs is almost imperceptible. The coloration of natural waters can vary depending on the characteristics and substances present (LIMA, 2001).

Turbidity does not depend strictly on the concentration of suspended sediments, but also on other characteristics of the sediment, such as size, mineral composition, color and amount of organic matter. Turbidity limits the penetration of sunlight, restricting photosynthesis, which in turn reduces oxygen replenishment (BRANCO, 1993).

High turbidity can influence aquatic communities, since it reduces the photosynthesis of submerged vegetation and algae, causing a suppression of fish productivity (CREPALLI, 2007).

Water temperature is a direct function of the speed of chemical reactions, the absorption of oxygen, the precipitation of compounds, the solubility of substances and the metabolism of organisms in the aquatic environment. When it is slightly elevated, it results in the loss of gases through the water, generating odors and ecological imbalance (VON SPERLING, 2005). It can be influenced by natural and anthropogenic factors. Natural factors generally come from the region's climatic regime and anthropogenic factors mainly from industrial waste and cooling water from machinery and boilers (BÁRBARA, 2006 *apud* SANTOS, 2009).

It is a parameter of fundamental importance because the concentration of dissolved oxygen depends directly on water temperature, which can affect aquatic biota. (2000) increases in temperature result in a reduction in dissolved oxygen and oxygen consumption due to the stimulation of biological activities.

In addition, temperature is a determining factor in the speed of a series of reactions that affect the chemical, physical and biological processes of the aquatic environment (GLEBER, 2002).

Electrical conductivity is the capacity of water to transmit an electric current. It is expressed in microSiemens.$_{cm-1}$ (p.S.$_{cm-1}$) and occurs due to the presence of dissolved substances. This property is directly related to the temperature of the body of water and the concentration of ionic substances dissolved in it (MACIEL JR., 2000). Some factors can influence the ionic composition of water bodies, such as the geology of the basin and the rainfall regime (LIMA, 2001).

Each body of water has a relatively constant degree of conductivity, which can be used for comparison purposes with regular measurements, and any significant change can indicate pollution (GLEBER, 2002). According to Libanio (2005), natural waters have an electrical conductivity of less than 100 pS.cm^{-1} , and can reach 1000 p.S.cm^{-1} when receiving high effluent loads.

In this context, electrical conductivity is considered an indirect measure of pollution, as it can be used to quantify the macronutrients present in the aquatic environment, obtain information on the decomposition of organic matter, identify pollution sources and hydrogeochemical differences, among others (SARDINHA *et al*, 2008 *apud* SANTOS 2009).

When it comes to solids, the amount and nature of dissolved and non-dissolved matter that occurs in the liquid medium varies gradually. Branco (1993) points out that all water contaminants, with the exception of dissolved gases, contribute to the solids load, which can be classified by their physical (dissolved and suspended) and chemical (organic and inorganic) characteristics.

Dissolved solids are found naturally in water due to the weathering of rocks and in large concentrations as a result of the discharge of domestic and industrial effluents. These particles are formed by the concentration of cations, anions and salts resulting from the combination of these ions that are dissolved in the water and suspended materials. Excessive dissolved solids in water can cause taste alterations (GLEBER, 2002; MACIEL JR., 2000).

Suspended solids are divided into sedimentable and non-sedimentable and originate from soil being carried away by surface runoff due to erosive processes and deforestation in the basin, effluent runoff and dredging for sand removal and mining activities. High concentrations of these solids increase turbidity, impairing the productivity of aquatic biota, cause changes in the color and odor of the water, act as carriers of adsorbed toxic substances and, in reservoirs, accelerate the silting process and block water intake structures (GLEBER, 2002; MACIEL JR., 2000).

In terms of chemical characteristics, however, APHA (1999) points out that the determination of fixed and volatile solids does not distinguish exactly between organic and inorganic materials because the loss of weight due to heating is not limited to organic material, but also includes loss due to decomposition or volatilization of some mineral salts such as: carbonates, chlorides, sulphates, ammonium salts, among others.

In this context, the presence of solids of any kind in water causes a change in color, an increase in turbidity and a decrease in transparency, which can affect the aquatic ecosystem due to a decrease in photosynthetic production and, consequently, dissolved oxygen in the body of water (BÁRBARA, 2006 *apud* SANTOS, 2009).

Odor, on the other hand, is associated both with the presence of chemical substances or gases dissolved in the water and with the metabolism of certain microorganisms, such as algae and cyanobacteria (LIBÁNIO, 2005).

The hydrogen potential expresses the intensity of the acidic (H^+) or alkaline (OH^-) condition of a solution, in terms of the concentration of hydrogen ions. H^+ is defined as the negative logarithm of the molar concentration of hydrogen ions (LIMA, 2001).

$$pH = - \log [H+]$$

The pH varies from 0 to 14, with 7.0 being neutral; below 7.0 the water is considered acidic and above 7.0 it is alkaline (MACIEL JR., 2000).

This parameter is formed by the presence of dissolved solids and gases in the water resource from the dissolution of rocks, absorption of gases from the atmosphere, oxidation of organic matter, photosynthesis and, in particular, the discharge of effluents (SANTOS, 2009).

According to Libanio (2005), pH influences the degree of solubility of various substances, the distribution of free and ionized forms of various chemical compounds, as well as defining the toxicity potential of some elements. For example, very basic pH values (above 8.0) tend to solubilize toxic ammonia ($NH3$), heavy metals and other salts in water and precipitate carbonate salts. Very acidic pH values (below 6.0) tend to increase the concentration of carbon dioxide ($CO2$) and carbonic acid in the water ($H2\ CO3$), (CREPALLI, 2007).

Alkalinity represents the ability of an aqueous system to neutralize acids. Although many compounds can contribute to the increase of this constituent in water, the largest fraction is mainly due to bicarbonates. The alkalinity of water has no implications for public health and is only considered unpleasant to the taste. The variables alkalinity, pH and carbon dioxide content are related to each other in nature. pH is a measure of the hydrogen concentration of water or another solution, controlled by chemical reactions and the balance between the ions present.

It is essentially a function of dissolved carbon dioxide and the alkalinity of the water (FEITOSA *et al*, 1997).

High alkalinity values in water bodies are related to the decomposition of organic matter, the respiratory activity of microorganisms and the discharge of industrial effluents (LIBÂNIO, 2005).

Acidity, on the other hand, is the inverse of alkalinity and can be caused by the decomposition of organic matter, the discharge of industrial effluents and the leaching of soil from mining areas (LIBÂNIO, 2005).

Chemical Oxygen Demand is related to organic matter and its polluting potential. It is a measure of the amount of oxygen consumed by the chemical oxidation of organic substances present in water (VON SPERLING, 2005). The COD test is based on the fact that almost all organic compounds can be oxidized by the action of a strong oxidizing agent in an acidic environment (MELLO, 2006 *apud* SANTOS 2009).

Like BOD, high COD values come from domestic and industrial effluents or water leached from animal farms (LIBÂNIO, 2005).

Phosphorus is an indispensable nutrient for all forms of life, since it participates in the processes of respiration, photosynthesis and cell reproduction. It is also extremely important for the growth of microorganisms that stabilize the organic matter present in water (MACIEL JR., 2000).

According to Feitosa *et al.* (1997), phosphorus is found in natural and wastewater almost exclusively in the form of phosphate. Due to the action of microorganisms, the concentration of phosphorus can be low (< 0.5 mg/l) in natural waters and values above 1.0 mg/l are generally indicative of polluted waters.

In the aquatic environment, phosphorus is present as organic phosphate and inorganic phosphate, distributed mainly in the forms of dissolved orthophosphates and organically bound phosphates (GLEBER, 2002). According to Crepalli (2007), high concentrations of phosphate come from domestic and industrial effluents, detergents, animal excrement and agricultural fertilizers.

The high concentration of phosphorus leads to excessive proliferation of algae and consequent eutrophication of the water body, causing alterations in the physical and chemical conditions of the water and in the aquatic community (MACIEL JR., 2000).

Nitrogen, one of the most important elements in the metabolism of aquatic ecosystems, has a complex chemistry, due to the various stages it can take on, and the impacts that changing its oxidation state can have on living organisms. A phenomenon best understood through the study of the nitrogen cycle describes these stages, emphasizing that the atmosphere serves as a reservoir in which nitrogen is constantly renewed by the action of electrical discharges and the fixation of bacteria. During these discharges, large quantities of nitrogen are oxidized to N2O5 and their union with water produces HNO3, which is normally carried to earth in the rain. Nitrates are also produced by the direct oxidation of nitrogen or ammonia and are also found in commercial fertilizers (LIMA 2001).

The forms in which this element is found in nature are: ammonia (NH_3), nitrite (NO2), nitrate (NO3$^-$), ammonium ion (NH4$^+$), molecular nitrogen (N_2), nitrous oxide (N_2O), dissolved organic nitrogen and particulate organic nitrogen, and determining the predominant form of nitrogen can provide indications of the stage of pollution. Therefore, if the pollution is recent, the nitrogen will basically be in the form of organic nitrogen or ammonia, since their oxidation has not occurred, and if it is old, it will basically be in the form of nitrate (VON SPERLING, 2005).

As nitrite is a reductive intermediate and is rapidly oxidized to nitrate, its concentration is generally low (1 mg/L). Nitrate is formed in the effluent by the action of microorganisms and by the chemical oxidation of ammonia. The older the effluent, the higher the nitrite content and the lower the organic nitrogen content. The COD/NKJT ratio is approximately 20 and can vary between 10 and 30 (RECESA, 2008).

The nitrogen cycle in natural waters is directly related to the level of dissolved oxygen in the body of water. Therefore, changes in the concentration of nitrogen in the water environment can lead to a series of problems with other water quality parameters. Basically, the problems caused by nitrogen are the result of the processes of nitrification/denitrification and eutrophication, as well as nitrate pollution and the high concentration of toxic ammonia present in the water (CHAPRA, 1997 *apud* SANTOS, 2009).

Dissolved oxygen is the most important parameter for expressing the quality of an aquatic environment, since it is essential for the maintenance of aerobic aquatic organisms (LIBÁNIO, 2005; MACIEL JR., 2000). In this way, the aquatic environment produces and consumes oxygen, which is taken from the atmosphere by the water/air interface and the photosynthetic processes of algae and plants (GLEBER, 2002; RECESA, 2008).

DO levels fluctuate seasonally and over 24-hour periods. Normally, natural waters have a concentration of around 8.0 mg/L at 25 °C, while the minimum concentration for the maintenance of aquatic biota is between 2.0 mg/L and 5.0 mg/L (LIBÁNIO, 2005).

This concentration can be reduced by the leaching of organic waste, through the consumption of DO by microorganisms in their metabolic processes of utilization and stabilization of organic matter (VON SPERLING, 2005). Just as the concentration of DO can be reduced, it can also be saturated. That is, DO saturations can come from photosynthetic processes, indicating eutrophication of the aquatic system. Eutrophied waters can have DO concentrations higher than 10 mg/L, even at temperatures above 20 °C (CREPALLI, 2007). In this sense, the determination of DO is fundamental for assessing the natural conditions of a river's water and detecting environmental impacts on it (MELLO, 2006 *apud* SANTOS, 2009).

According to FEPAM (2010), the factors that most influence the concentration of this gas in the aquatic environment are: water temperature (the higher it is, the lower the concentration of DO present in the water), atmospheric pressure (altitude) and salinity.

Biochemical oxygen demand reflects the amount of oxygen required to stabilize carbonaceous organic matter through biochemical processes. It is an indirect indication of biodegradable organic carbon. Stabilization takes about 20 days for domestic sewage, which corresponds to the final BOD. To avoid this delay, the standard BOD was standardized, which determines the test to be carried out after the fifth day of consumption at a temperature of 20 °C (APHA, 1999).

According to Eiger (2003), BOD has two major advantages: it makes it possible to compare the polluting potential of different effluents from a wide variety of sources according to the same scale and to assess the state of water quality in any given river, since it is an indirect measure of the consumption of dissolved oxygen in the water environment.

Just like COD, high BOD values come from domestic and industrial effluents or water leached from animal farms (LIBÂNIO, 2005).

Oils and greases comprise a group of substances involving oils, greases, waxes and fatty acids, which are present in leftover butter, margarine, vegetable fats and oils, red meat fats, as well as a portion of oily matter due to the presence of lubricants used in industrial establishments (GUIMARÂES *et al* , 2002). These substances are extracted using a solvent (extractant) recommended by the experimental method in APHA (1999). The solvent extraction method makes it possible to separate the oils and greases from the solid and liquid phases of the effluent through the evaporation process.

The importance of determining the oil and grease content (TOG) is due to the fact that when high concentrations of oil and grease are present in wastewater, they cause operational problems at the primary treatment stage and can interfere with biological (secondary) treatment. These problems are caused by the fact that oils and greases resist anaerobic digestion, causing accumulations of scum in the digesters and making it impossible to use the sludge for fertilization (GUIMARÂES *et al*, 2002).

Jordao and Pessôa (1995) point out that the need to remove grease is to prevent clogging of collectors as well as adherence to special parts of the sewage system and, above all, accumulation in treatment units, since oils and grease cause unpleasant odors and disturb treatment devices.

Surfactants are molecules made up of a hydrophobic part and a hydrophilic part. The apolar part of the molecule is often a hydrocarbon chain while the polar part can be ionic (anionic or cationic), non-ionic or amphoteric. Most commercially available surfactants are synthesized from petroleum derivatives (Nitschke & Pastore, 2002).

Heavy metals are rarely found in natural waters, and concentrations of these elements in water bodies generally come from the discharge of industrial effluents and leaching from mining areas (LIBÂNIO, 2005).

The most toxic metals are: aluminum (Al), copper (Cu), chromium (Cr), tin (Sn), nickel (Ni), mercury (Hg), vanadium (V) and zinc (Zn). These have high bioaccumulation factors, since they are considered substances that are preserved in the system, even if transformations, sedimentation and/or resolubilization occur (GIORDANO, 2005).

Coliforms are indicators that the water body is contaminated by domestic sewage, since this group of bacteria inhabits the intestinal tract of humans and animals (VON SPERLING, 2005).

Although this group of bacteria are mostly non-pathogenic, they serve as indicators of potential contamination by pathogenic bacteria, viruses and protozoa that also reside in the intestinal tract. In addition, coliforms also occur in smaller quantities in natural environments, such as pastures, soils and submerged plants, and are therefore referred to as total coliforms. While fecal coliforms are bacteria specific to the intestinal tract (GLEBER, 2002).

Chapter 4

4 MATERIAL AND METHODS

4.1 Study area

The public slaughterhouse in the municipality of Pombal-PB is located in the rural area of the city, to the west of the state of Paraíba, more precisely at the geographical coordinates of latitude 6°44'41.61"S and longitude 37°44'57.61"O, as shown in Figure 2 below. The municipality is located in the hydrographic region of the Piranhas River to the south.

The aim of this research was to provide an overview of how the issue of animal slaughter has been dealt with in the municipality of Pombal -PB.

FIGURE 2 - Location of the public slaughterhouse in the municipality of Pombal-PB.

Source - Google Earth images from 10/03/2016

4.2 Wastewater treatment system equipment

In order to characterize the slaughterhouse's effluent treatment system and its equipment, *on-site* visits, photos, measurements and the layout of the Effluent Treatment System (STE) were carried out. In order to assess the efficiency of this system, physical-chemical analyses were carried out to determine the quality parameters.

4.3 Effluent collected and collection points

In order to characterize the STE effluent and carry out the research, two main points/locations were defined as references for the collection of samples: Point 1 secondary treatment (after grating), point 2 after tertiary treatment (after skeptic fume), these were defined as the points that best characterize the effluent for subsequent suggested reuse.

Samples were taken from August 2015 to January 2016; four samples were taken on different dates and in different months, and on each date 2L samples were taken from each point, always in the morning (slaughter time).

All the samples were of the composite type, for better representation of the effluent, where they consisted of 500 ml of effluent collected at intervals of one hour and one hour for a period of 4 hours, and during collection the effluent was stored in sterilized polypropylene containers, in accordance with Brazilian standard NBR 9898/1987. After the composite samples were taken, the effluent was homogenized and stored in small 2-litre containers. At the time of each collection, readings were taken of certain parameters such as pH and effluent temperature, ambient temperature and flow measurement.

Flow rates were obtained through mechanical measurements, with stopwatches and graduated containers (graduated beakers), obtaining the volume/time ratio.

4.4 Physico-chemical characterization of the effluent samples collected

The physicochemical characterization of the effluent samples collected at the two points of the effluent treatment system of the slaughterhouse in the municipality of Pombal-PB was carried out with regard to the following parameters: Total Phosphorus (TP), Total Nitrogen (TN), Chemical Oxygen Demand (COD), Biochemical Oxygen Demand (BOD), Sedimentable Solids (SSed), Oils and Grease, Turbidity and pH. The analytical methods used to characterize the effluent samples are official and are described in the Standard Methods for the Examination of Water and Wastewater (APHA et al., 2012).

The samples collected were analyzed at the Biotechnology, Environmental Sanitation and

Water and Effluent Quality Laboratory of the IFPB - Federal Institute of Paraíba, Souza campus. All the analyses were carried out at a controlled temperature of 20° C in triplicate and the results are expressed as an average of the values found. These parameters were selected because they best characterize the STE effluent, for comparison with CONSEMA 128/06 and CONAMA 357/05.

4.5 Application of reuse water

After analyzing the physical and chemical parameters, a comparative study was carried out between the results obtained and the standards required by current Brazilian legislation and the desired rates for reusing the effluent.

4.6 Designing an Effluent Treatment Plant

A pilot project was carried out for an effluent treatment plant that would satisfactorily serve the municipal slaughterhouse in Pombal-PB.

4.7 Statistical analysis

The ASSISTAT program version 7.5 Beta (Azevedo, 2009) was used to analyze the data. Tukey's 5% probability test was used to check for possible statistical differences between the parameters determined in the effluent.

Chapter 5

5 RESULTS AND DISCUSSION

5.1 Characterization of the slaughter process

The results of the characterization of the slaughter process depend on the characteristics of the region in which the slaughterhouse is located. According to data provided by city officials, the maximum daily slaughter capacity is 50 cattle and 200 pigs. The water consumption for cattle is approximately 1000 liters per head and for pigs 700 liters per head per day.

According to the data provided by the slaughterhouse under study, approximately 200 pigs and 80 cattle are slaughtered each week, with an average water consumption of 700 liters per pig and 1000 liters per cattle slaughtered, generating around 94 cubic meters of liquid effluent each week. It has approximately 12 employees, who contribute 15 cubic meters of sanitary sewage per week, which after treatment (septic tank), goes into an adsorption ditch (Infiltrado).

The volume of water used comes from the municipality's water tankers (Figure 3), which have to be supplied every slaughter day.

FIGURE 3: Water supply to the slaughterhouse from water tankers

The slaughterhouse's industrial process basically consists of the following stages: receiving platform, sties, stunning, bleeding, scalding and depilation; the entire process is carried out manually.

5.2 Characterization of STE

Liquid waste from the pig process is generated by bathing the animals; washing the floors of the stunning pens and other areas of the production process (vomiting, bleeding, skinning and sanitizing the animal's cavities so that waste is perfectly removed). The washing of the floors of the stunning boxes (pens) is proportional to the number of animals received, as there is no constant number of washes. Washing takes place according to the number of animals received, but the more animals received, the greater the number of washes. At the end of the day, a wash is carried out with the addition of hypochlorite to disinfect the area. All the employees carry out this procedure, but none have had specific training to carry out this activity.

The resulting effluent load is always constant, as the load is the result of concentration and flow rate.

The slaughterhouse under study has preliminary treatment through sieving, a physical-chemical system through decanting, and biological treatment through septic tanks.

Evaluating the treatment system in place at the Pombal-PB public slaughterhouse, it is clear that it needs to be altered, given that the system consists only of a septic tank and stabilization ponds (Figure 4 below), with no adequate structure to mitigate the effluents generated.

The effluent treatment system of the slaughterhouse in the municipality of Pombal -PB is carried out routinely, without any type of quality control in its documentary form (registration, analysis, etc.), and it does not have an environmental license.

Figure 4: Stabilization ponds

Source: Author's archive

5.2.1 Parameters obtained

A) Flow rate

As there is no license issued by the competent body, it is difficult to know the operating capacity of the slaughterhouse, however the construction project states that it would be able to flow up to 150 m³ per day.

As there are no flow measuring instruments in the slaughterhouse, quantification was carried out manually using a stopwatch, where a container with a known volume (graduated beakers) was taken and the time taken to fill it completely was calculated.

The results of the measurement analyses showed average values of 31.75 m³ /day at collection point 1 and average values of 31.35 m³ /day at collection point 2. These values show that the slaughterhouse operates within the limits of the maximum operating flows.

B) Hydrogen Potential (pH)

The resulting average pH values remained constant throughout the treatment system, with values varying little between 8.19 for point 1 and 7.63 for point 2 where the samples were taken. As can be seen in Figure 14, these values are within the range of 6.0 to 9.0 established by CONAMA Resolution 357/05 and CONSEMA Resolution 128/06.

C) Chemical Oxygen Demand (COD)

The results obtained for COD at the collection points ranged from 3353 mg/L to 2197 mg/L at collection points 1 and 2, respectively. These average values are well above the limits required by CONAMA 357/05 legislation, which establishes values of < 120 mg/L for slaughterhouse effluent.

D) Biochemical Oxygen Demand (BOD)

According to CONAMA Resolution 357/05 and CONSEMA 128/06, maximum concentrations of 350 mg/L are set for Biochemical Oxygen Demand (BOD). The results of the analysis showed average values of 1378 mg/L for point 1 and average values of 894 mg/L for point 2 where the effluent samples were collected. These values are well above the values stipulated by the aforementioned resolutions.

The liquid effluents generated by the slaughtering industry generally have high concentrations of biochemical oxygen demand (BOD), chemical oxygen demand (COD), suspended solids, oils and grease (OG) and organic nitrogen. The blood secreted into the effluent is the major factor behind the high BOD and COD levels in these effluents.

E) Phosphorus

The total phosphorus concentrations resulting from the samples taken at collection points 1 and 2, respectively, ranged from 19.3 mg/L to 17.6 mg/L. The values established by the legislation are <3 mg/L, so these are well above the required values stipulated by the legislation.

F) Nitrogen

The average values found for the concentration of total nitrogen varied from 35 mg/L to 56 mg/L for the 2 and 1 samples analyzed. These nitrogen concentration values are above the limit stipulated by CONAMA 357/05, which establishes values of <20 mg/L.

G) Sedimentable solids

The average values for settleable solids throughout the treatment system were 37.0 mg/L for point 1 and 22 mg/L for point 2. These values are outside the limits set by CONAMA Resolution 357/05 of <1 mg/L.

H) Oil and grease

The average values obtained from the two collection points for oil and grease levels showed levels of 400 mg/L for collection point 1 and 289 mg/L for collection point 2, which are above the standards stipulated by the legislation of <30 mg/L.

I) Turbidity

The results obtained for turbidity at the collection points ranged from **331** NTU to **267** NTU - (Number of Turbidity Units), according to CONAMA 357/05 the standards are not up to the required 100 NTU.

J) Temperature

It can be seen that the effluent temperatures obtained on the sampling days were below the maximum limits considered by CONSEMA 128/06 legislation of <40 °C, with average values of 39.2 °C and 29.9 °C respectively for sampling point 1 and point 2.

5.3 Statistical analysis

Table 4 shows all the parameters analyzed in the effluent from the public slaughterhouse in the municipality of Pombal -PB, where the Tukey mean comparison test at 5% probability was applied to determine the statistical differences.

TABLE 4 - Average values of the physicochemical characterization of the parameters analysis of the effluent from the public slaughterhouse in the municipality of Pombal -PB

Analyzed slaughterhouse effluent parameters	Collection point 1	Collection point 2	Parámetros legislado CONAMA n° 357/05 e CONSEMA 128/06
Flow (m^3 /day)	31,75 a	31,35 a	m /day^3

pH	8,19 a	7,63 b	6,0 a 9,0
COD (mg/L)	3353,0 a	2197,0 b	< 120 (mg/L
BOD (mg/L)	1378,0 a	894,0 b	< 350 (mg/L
Phosphorus (mg/L)	19,3 a	17,6 b	< 3.0 (mg/L
Nitrogen (mg/L)	56,0 a	35,0 b	< 20.0 (mg/L
Sed. solids (mg/L)	37,0 a	22,0 b	< 1.0 (mg/L
Oils and Greases (mg/L)	400,0 a	289,0 b	< 30.0 (mg/L
Turbidity (NTU)	331,0 a	267,0 b	100 NTU
Temperature (°C)	39,2 a	29,9 b	< 40 °C

Averages followed by different lowercase letters in the rows differ according to Tukey's test ($p < 0.05$).

Chapter 6

6 - Recommended Treatment and Indication of Reuse

A project was drawn up for an effluent treatment plant that would satisfactorily serve the municipal slaughterhouse of Pombal-PB, with the principle of adapting it to the layout of the existing treatment plant, which will be presented to the municipal public managers for future feasibility.

6.1 - Suggested equipment to build the new WWTP in order to optimize the process are:

6.1.1- Railing

The purpose of this unit operation is to retain suspended coarse solids in the wastewater, preventing them from passing through to the other devices in the WWTP.

6.1.2- Flow meter - Parshal flume

A Parshal flume should be installed at the entrance to the unit to monitor the operational flow of effluent and any other adjustments that may be necessary.

6.1.3- Grease trap / sand box.

Their basic function is to separate the fats contained in liquid effluents. The tanks are also used as sandboxes. The project provides for two separate channels for individual operations. This will allow one tank to be cleaned while the other is in operation.

6.1.4- Equalization / Aerated Tanks

The purpose of equalization / aeration tanks is to regulate flow, pH, temperature, turbidity, solids, BOD, COD and color and, in addition, to aerate the entire mass of effluent from the industrial process.

The tank must remain at least 1.0m deep so that the effluent flow is always homogenized and the pumps do not run dry.

Perfect homogenization of the liquid mass will be achieved by the action of the floating aerator which, as already mentioned, will homogenize the contents of the tank.

6.1.5- Decant tanks

After being subjected to aeration in the previous stage of treatment, the effluent will be treated with chemical flocculation agents in order to promote the sedimentation of materials that were previously dispersed in the fluid mass. Forwarding the effluent to the decanter will make this removal possible. The denser material will be deposited at the bottom of the equipment while the fluid phase free of these solid materials will be sent to the next stage - Chlorination.

6.1.6- -Treated Effluent Reception Pool

This pond will receive the treated effluent and pump it either to the dispersal system corresponding to the "Nitrification Field", which will receive the treated effluent and finally dispose of it in the soil, or to the irrigation system, which will supply water to the areas (gardens) previously defined by the company for this purpose.

6.1.7- -Absorption range (Infiltration)

The effluent will be sent to soil infiltration chambers, as an alternative final destination for the portion of the effluent that has not been used in non-noble washing and gardening operations. Technical details are attached.

6.2 - Reuse application

The reuse of treated effluents is an important alternative due to the growing scarcity of water resources, especially for the agro-industry in this study. Based on an analysis of the local reality in which the slaughterhouse is located and taking into account economic, social and, above all, environmental factors, it is proposed that the waste after the indicated treatment can be reused in the following ways:

a) Washing the sidewalks and rooms of the public slaughterhouse under study.

b) Irrigation of green areas such as squares, gardens, flowerbeds, etc.

c) Washing official municipal vehicles.

CONCLUSIONS

According to the results of this study, we can conclude that:

The effluent from the treatment system of the municipal slaughterhouse in Pombal - PB is in disagreement with the parameters stipulated by current Brazilian legislation, thus violating the environmental laws of this activity.

Among the physicochemical characteristics analyzed in the effluent from the slaughterhouse treatment system, the BOD and COD parameters stand out, which showed very high levels when compared to the parameters stipulated by the legislation, these being indicators of the efficiency of the effluent treatment, it is clear the need to implement corrective measures to improve the effluent treatment process.

It was possible to design a suitable treatment system for this activity, suggesting modifications to the existing local effluent treatment plant in order to optimize the process and meet the requirements of the relevant environmental legislation.

Once the modifications to the effluent treatment system at the Pombal - PB municipal slaughterhouse suggested in this work have been carried out, it will be possible to implement techniques for reusing this effluent within the slaughterhouse itself, thus promoting the preservation of water resources, treated water for less noble purposes and reducing the environmental impacts of this activity.

SUGGESTIONS

Enabling Pombal's municipal government and its environment secretary to commit public managers to public policies aimed at complying with legislation on adapting the municipal slaughterhouse's treatment system and reusing this effluent in order to promote the preservation of local water resources and the environment.

Implement routine practices for evaluating the Water Quality Index of the municipal slaughterhouse treatment system and reuse, systematically evaluating the process.

BIBLIOGRAPHICAL REFERENCES

ABIPECS - **Brazilian Pork Producing and Exporting Industry Association.** Available at: http://www.abipecs.org.br/uploads/relatorios/mercadoexterno/exportacoes/anuais/jan-dez- 2014_jan-dez-2013.pdf>. Accessed on: 10 Feb. 2015.

ABIEC, **Associação Brasileira das Indústrias Exportadoras de Carne** - ABIEC (2014). Available at http://www.abiec.com.br/notícias/export-carne. Accessed on January 3, 20015.

ASANO, T. **"Wastewater reuse cuts down waste".** Water Quality International. N° 2, 1998.

BEZERRA, L. F; MATSUMOTO, T. Evaluation of the removal of carbonaceous and nitrogenous organic matter from wastewater in a membrane bioreactor. **Revista Engenharia Sanitária e Ambiental.** [online]. 2011, vol.16, n.3, pp. 253-260. Available at: <http://www.scielo.br/pdf/esa/v16n3/v16n3a08.pdf>. Accessed on: August 5, 2013.

BLUM, J. R. C. **Water quality criteria and standards.** In: Mancuso, P. C. S; Santos, H. F. Reúso de água. Chap. 5, Barueri, São Paulo: Manoli, 2003.

BRANCO, Otavio Eurico de Aquino. **Water Availability Assessment: concepts and applicability.** São Luis - MA, 2006;

BRAZIL, National Water Resources Council - CNRH. Resolution No. 54, of November 28, 2005. Establishes modalities, guidelines and general criteria for the practice of direct non-potable water reuse, and other measures. **Official Gazette of the Federative Republic of Brazil**, Brasília, DF, November 28, 2005

BRAZIL, National Environmental Council - CONAMA. Resolution No. 357, of March 17, 2005. Provides for the classification of bodies of water and environmental guidelines for their classification, as well as establishing the conditions and standards for the discharge of effluents, and makes other provisions. **Official Gazette of the Federative Republic of Brazil**, Brasilia, DF, March 17, 2005.

BRAZIL. Federal Law No. 9.984, of July 17, 2000. Provides for the creation of the National Water Agency - ANA, a federal entity for implementing the National Water Resources Policy and coordinating the National Water Resources Management System, and makes other provisions. **Official Gazette of the Federative Republic of Brazil, Brasília**, DF, July 17, 2000.

CHILE, CONAMA - NATIONAL ENVIRONMENTAL COMMISSION. Guide for the Control and Prevention of Industrial Contamination - **Meat Processing Industry**, 1998.

CREPALLI, M. S. **Water quality of the Cascavel River.** Dissertation (Postgraduate Program in Agricultural Engineering) - State University of Western Paraná, Paraná, 2007.

EPA - **Environmental Protection Agency. Guidelines for Water Reuse.** EPA/625/R04/108, Washington, DC, September, 2004. Available at: <http://nepis.epa.gov/Adobe/PDF/30006MKD.pdf>. Accessed on: 02 Apr 2014.

ENVIROWISE, U. K. **Environmental good practice guide: Reducing water and effluent costs in red meat abattoirs.** GG234. 2000. Available at: <http://infohouse.p2ric.org/ref/23/22904.pdf>. Accessed on: 30 Dec 2014.

FAPPI, D. A. **Micro and ultrafiltration as post-treatment for the reuse of effluents from pig slaughterhouses and slaughterhouses.** Dissertation (Master's Degree) - Federal Technological University of Paraná. Postgraduate Program in Environmental Technologies. Medianeira, 2015.

FIESP - **Federation of Industries of the State of São Paulo. Manual for water conservation and reuse in industry.** Rio de Janeiro, 2006. 32p. Available at:<http://www.firjan.org.br/lumis/portal/file/fileDownload.jsp?fileId=4028808120E98

EC70121222C66745337>. Accessed on September 3, 2013.
FILHO, D. B. MANCUSO, P. C. S. **Concept of water reuse. Water reuse**. In:
Mancuso, P.C.S; Santos, H. F. Reúso de água. Chap. 2, Barueri, Sao Paulo: Manoli,
2003.
FRACACIO, N. **Water use in industrial activities**. Sao Paulo, 2009. Available at:
<http://prope.unesp.br/xxi_cic/27_36876131866.pdf>. Accessed on September 5, 2013.
GLEBER, L. **Reducing the risks of environmental impact in integrated vegetable
production**. Technical circular, n. 38, July/2002.
HESPANHOL, I. Water reuse potential in Brazil: agriculture, industry, municipalities,
aquifer recharge. **Revista Bahia Análise e Dados,** Salvador, v. 13, n. especial, 2013.
JORDÄO, E. P.; PESSOA, C. A. Tratamento de esgotos domésticos. 3 ed. Rio de
Janeiro: **ABES**, 1995. 692 p.
HAMMER, M. J; HAMMER, M. J. **Water and wastewater technology**. Ed. Pearson
Prentice Hall. 6 th ed. 553 p, 2007.
KRIEGER, E. I. F. **Evaluation of water consumption, rationalized use and reuse of
liquid effluent from a pig slaughterhouse in the search for the company's socio-
environmental sustainability**. 2007. 130 p. Thesis (Doctorate in Ecology). Federal
University of Rio Grande do   Sul   , Porto Alegre, 2007.
Available at:<www.lume.ufrgs.br/bitstream/handle/10183/12050/000618507.pdf?...
Accessed on: July 15, 2014.
KRIEGER, E. I. F; RODRIGUEZ, M. T. R. **Water balance in a pig slaughterhouse
and evaluation of water use in holding pens**. 2007. Available at:
<http://www.bvsde.paho.org/bvsaidis/uruguay30/BR10559_Krieger.pdf>. Accessed on:
Feb. 10, 2015.
LEMOS, C. A. **Water** quality in a **watershed in the Atlantic Forest Biosphere
Reserve, Maquiné, Rio Grande do Sul, Brazil**. Dissertation (Postgraduate Program in
Ecology - Institute of Biosciences). Federal University of Rio Grande do Sul, 2003.
LENNTECH, **Water Treatment Solutions**. Characteristics of boiler feed water. 2015.
Available at: <
http://www.lenntech.com.pt/aplicacoes/processo/caldeira/caldeira-agua-
alimentacaocaracteristicas.htm>. Accessed on October 30, 2015.
LEVINE, D. A. & ASANO, Water Reclamation, recycling and reuse in industry. In:
**Water recycling and resource recovery in industry: Analysis, technologies and
implementation.** IWA Publishing, 2002.
LIMA, E. B. N. R. **Integrated modeling for water quality management in the
Cuiabá River basin.** PhD Thesis in Civil Engineering - Federal University of Rio de
Janeiro, COPPE. Rio de Janeiro, 2001.
LOBO, L. P. **Comparative analysis of membrane separation and physical-chemical
clarification processes for water reuse in industry**. 2004. 108 p. Dissertation
(Master's Degree in Environmental Engineering). Rio de Janeiro State University, Rio
de Janeiro, 2004. Available at:
<http://www.peamb.eng.uerj.br/trabalhosconclusao/2004/lucianapaulalobopeamb200
4.pdf>. Accessed on: 29 Jan 2015.
LUIZ, D. B. **Combination of chemical treatments for the potabilization of water
discharged from slaughterhouses**. 2010. 185 p. Thesis (Doctorate in Chemical
Engineering). Federal University of Santa Catarina, Florianópolis, 2010. Available
at:<http://repositorio.ufsc.br/bitstream/handle/123456789/93618/285734.pdf?seque
nce=1&isAllowed=y>. Accessed on: 06 Jan 2015.
MIERZWA, J. C. **Rational use and reuse as a tool for water and effluent
management in industry - a case study of Kodak Brasileira**. 2002. Thesis (Hydraulic
and Sanitary Engineering), Polytechnic School of the University of São Paulo, 2002.
MANCUSO, P.C; SANTOS, H.F. **Reúso de Água**. Editora Manole Ltda, 576 p., 2003.

METCALF; EDDY. **Wastewater engineering treatment and reuse**. Mc Graw-Hill: Higler Education. Fourth Edition, 2003. 1819p

NOSCHANG, M. C. S. **Water management and reuse in agribusiness**. 2011. 107 p. Dissertation (Master's in Environmental Quality) Feevale University: Rio Grande do Sul, 2011. Available at: <http://ged.feevale.br/bibvirtual/Dissertacao/DissertacaoMarilizNoschang.pdf>. Accessed on Aug. 23, 2013.

NUNES, D. G. **Modeling self-depuration and water quality of the Turvo Sujo River**. Dissertation (Postgraduate Program in Agricultural Engineering). Federal University of Vigosa 2008.

OENNING JUNIOR, A. PAWLOWSKY, U. Evaluation of advanced technologies for water reuse in the metal-mechanical industry. **Eng. Sanit. Ambient.** [online]. 2007, vol.12, n.3, pp. 305-316. ISSN 1413-4152. Available at: <http://www.scielo.br/pdf/esa/v12n3/a08v12n3.pdf>. Accessed on 05 Aug 2013.

OLIVEIRA. M. D. **Development of a computerized system for environmental impact assessment**. Dissertation (Postgraduate Program in Environmental Engineering). Federal University of Espírito Santo. 2004.

PALUDO, José Roberto; BORBA, Julian. **Water supply and sanitation: a comparative study of management models in Santa Catarina**. Ambiente. soc., Sao Paulo, v. 16, n. 1, Mar. 2013. Available at:<http://www.scielo.br/scielo.php?script=sci_arttext&pid=

PINHO, F; VASCONCELOS, A.K; MARINHO, G. **Diagnosis of reuse in the Brazilian Northeast**. Paper presented at the III Research and Innovation Congress of the North-East Technological Education Network, Fortaleza, 2008. Available at: < http://www.intv.cefetce.br/connepi/papers/>. Accessed on September 20, 2010.

SALES, Luís Gustavo de. **Indicadores de Sustentabilidade Hidroambiental para Bacias Hidrográficas do Semiárido Brasileiro: uma proposta** de operacionalizagao **na sub-bacia do Rio do Peixe - PB**. Thesis (Doctorate in Natural Resources - UFCG), Campiña Grande - PB, 2014;

SANTOS, V. R., **Evaluation of the water quality of the Andrada River using the QUAL2k model**. Course Conclusion Paper - Environmental Engineering. University of Passo Fundo, 2009.

SANTOS, V. R., **Evaluation of the water quality of the Andrada River using the QUAL2k model**. Course Conclusion Paper - Environmental Engineering. University of Passo Fundo, 2009.

SENAI - NATIONAL SERVICE FOR INDUSTRY. PORTO ALEGRE. **Basic principles of cleaner production in slaughterhouses**. Cleaner Production Manual Series, 2003.

SEAB PR - **PARANÁ STATE SECRETARIAT OF AGRICULTURE AND SUPPLY**. Pig farming - Analysis of the agricultural situation. Paraná, 2013. Available at:<http://www.agricultura.pr.gov.br/arquivos/File/deral/Prognosticos/SuinoCultura_20 12 _2013.pdf>. Accessed on Aug. 20, 2014.

SCARASSATI, D; CARVALHO, R. F; DELGADO, V. 1; CONEGLIAN, C. M. R; BRITO, N. N; TONSO, S; SOBRINHO, G. D; PELEGRINI, R. **Treatment of effluents from slaughterhouses and meatpacking plants.** In III Fórum de Estudos Contábeis, [online], Claretianas, 2013. Available at: <http://www.ctec.ufal.br/professor/elca/TRATAMENTO%20DE%20EFLUENTES%20 DE%20MATADOUROS%20E%20FRIGOR%C3%8DFICOS.pdf>. Accessed on: Sep 25, 2015.

VON SPERLING, M. **Introduction to water quality and sewage treatment.** In: Princípios do tratamento biológico de águas residuais. 3 ed. Vol.1. Belo Horizonte: Department of Sanitary and Environmental Engineering, UFMG, 2005.

MIX
Papier aus verantwortungsvollen Quellen
Paper from responsible sources
FSC® C105338
FSC
www.fsc.org

Printed by Books on Demand GmbH, Norderstedt / Germany